IMAGES
of America

LONG ISLAND
AIRCRAFT CRASHES
1909–1959

IMAGES
of America

LONG ISLAND
AIRCRAFT CRASHES
1909–1959

Joshua Stoff

ARCADIA
PUBLISHING

Published by Arcadia Publishing
Charleston, South Carolina

Library of Congress Catalog Card Number: 2003116337

For all general information, contact Arcadia Publishing:
Telephone 843-853-2070
Fax 843-853-0044
E-mail sales@arcadiapublishing.com
For customer service and orders:
Toll-free 1-888-313-2665

Visit us on the Internet at www.arcadiapublishing.com

All photographs contained in this book are from the collection of the Cradle of Aviation Museum in Garden City, New York.

CONTENTS

A Crash, c. 1930. Seen is a fatal crash of a Fleet Model 2, near Roosevelt Field.

INTRODUCTION

During the first half-century of flight by man, Long Island, New York, was truly one of the centers of the aviation world. There were more aircraft manufacturers here and more airports located here than in any other part of the United States. Due to this extraordinary volume of air traffic, it is only logical to assume that Long Island also probably led the country, if not the world, in aircraft crashes. Normally, when one writes of the history of aviation, it is usually a story of continually better aircraft, better technology, new records, and new heroes. This book portrays the darker side of the history of aviation, how we got from there to here, on a road paved with tragedy. It is a story that must be told. Because of the many crashes here, Long Island ultimately saw the earliest air-traffic control systems, airport lighting, aviation weather reports, paved runways, and professionally run flying schools. It is estimated that the crashes shown in this book are probably only about 10 percent of those that occurred on Long Island during this period.

The reasons for these accidents are many. Pilots, on occasion, make mistakes. Weather is unpredictable, especially in aviation's pre–World War II days, when weather reports were often inaccurate, infrequent, or did not exist at all. Unlike today's small aircraft, early planes often suffered engine failure, and in-flight structural failure was not unheard of. The flying speeds of early aircraft were also so low they were always on the edge of a stall or spin. Prior to World War II, air-traffic control was nonexistent. Thus, midair collisions in the crowded skies around Long Island airports happened with alarming frequency. Furthermore, when one examines the photographs of these pre–World War II airfields, one can clearly see flying fields in poor condition. Ruts and mud were the routine, and these conditions undoubtedly led to many planes nosing over, especially combined with the fact that the planes usually had no brakes.

Accustomed as we are now to safety in the air, when flying around the world is safer than driving to a nearby store, it is hard to keep in mind just how dangerous flying really was in its early days. For the first 30 years or so, nearly all airplanes were fragile-looking wooden structures braced by wires and covered with varnished fabric. Even the safest of early aircraft were dangerous in comparison to today's planes.

Man paid dearly in his quest to conquer the sky. Many lives and fortunes were lost. However, what was learned from these accidents has brought about most of the improvements in modern aviation. Flying has clearly learned from its mistakes. Aviation today is a remarkably safe way to get from Point A to Point B. Airline passengers are far safer than if they drove between the same two points. There is no doubt that flying is increasingly becoming safer in spite of the continuous increase in air traffic. Aviation's accident rate in the 1930s was appalling when compared to the present-day rate. Regardless of the many accounts of accidents in this book, flying by any comparison is now a relatively risk-free form of travel. People can be assured that lessons learned from these accidents have been applied in making aviation continually safer. This history helps to show how flying got to be as safe as it is today.

Since the first aircraft fatality (Lt. Thomas Selfridge in a Wright aircraft in 1908), news of air crashes has always been met by the public with a combination of concern, horror, and fascination. There has always been a strange attraction to air crashes, perhaps because they are now so rare.

With the historic photographs on the following pages, one can appreciate how safe flying has really become simply by seeing just how unsafe it once was.

One

THE PIONEER YEARS
1909–1914

With the invention of the airplane in 1903 and its continued development in Europe and America, one of man's greatest dreams—to fly—was finally realized. By 1909, the first tentative flights were made from the flat, open central area of Nassau County, New York, then known as the Hempstead Plains. This inaugurated aviation in an area where it was to remain the focus of intense activity for the next 50 years. Early aviators, who often built their own airplanes, competed for fame and fortune while flying shockingly frail craft made of wood, covered with fabric, and braced by wires. The airfields of the day used existing unimproved open fields for flying, and horse-racing tracks also proved popular. These spindly craft were light, so they were easily upset by wind gusts or could flip over upon hitting a rut, and engine and even structural failure were also common. Nonetheless, by 1914, man had firmly established himself in his new element—the air.

NASSAU BOULEVARD AIRFIELD, GARDEN CITY, 1911. One of three airfields in central Nassau County in the pre–World War I period, Nassau Boulevard was one of the best-developed airfields of its day. It featured some three dozen wooden sheds, a grandstand, a restaurant, offices, and a large, open, level flying field. Here, a Wright Model B banks over the crowd.

A WALDEN MODEL III, MINEOLA FLYING FIELD, DECEMBER 1909. The Walden III was the first American monoplane to fly and was built by Dr. Henry Walden, a Long Island dentist. It first flew on December 9, 1909.

A CRASH OF THE WALDEN MODEL III, AUGUST 3, 1910. Long Island's first aircraft crash occurred when the Walden III got caught in a downdraft at the edge of the Mineola Flying Field, seriously injuring Dr. Walden. This was only the second serious aircraft crash in the United States.

A CURTISS MODEL D PUSHER BIPLANE, 1911. A popular type of American aircraft in the pioneer era, the Curtiss D usually had a 60-horsepower Curtiss engine and flew at speeds around 50 miles per hour.

JAMES WEEKS, 1911. James Weeks is seen above in his Curtiss D aircraft. Below, he stands next to the wreck of his plane on the Hempstead Plains. When asked why he stopped flying, he said, "First, the money went out of it, second, I was spending too much time in hospitals."

A WRIGHT MODEL B PUSHER BIPLANE ON THE SHEEPSHEAD BAY RACETRACK, 1911. A four-cylinder Wright engine drove two chain-driven propellers, allowing this plane to reach speeds of 50 miles per hour.

THE WRECK OF A WRIGHT B ON THE HEMPSTEAD PLAINS, 1911. All Wright aircraft were generally unstable, and it was not uncommon for aircraft to spin in from altitude.

A BLERIOT TYPE XI AT THE BELMONT PARK AIR MEET, OCTOBER 1910. Fast for its day, the Bleriot was powered by an 80-horsepower rotary engine, allowing it to reach 75 miles per hour.

THE CRASH OF JOHN MOISANT'S BLERIOT AT THE BELMONT PARK MEET. John Moisant survived the crash and went on to win an important speed race. The crash was caused by his taxiing into another aircraft, a common occurrence on aircraft with no brakes or steerable wheels.

THE WRIGHT BABY GRAND CRASH AT THE BELMONT PARK MEET. Walter Brookins was the pilot of this airplane. The crash was caused by the loss of control at low altitude. If not for the accident, the plane probably would have won the speed race.

THE CRASH OF THE HEINRICH MODEL D ON THE HEMPSTEAD PLAINS, 1912. A locally built aircraft, the Heinrich D was a rather streamlined design for its day. The Heinrich brothers built several successful aircraft through 1917.

16

Two

WORLD WAR I
1914–1918

Little more than a decade after the Wright brothers first flew, the airplane changed the world forever; it had become a weapon of war. Planes first served as observation platforms, but machine guns and bombs were soon added, and the fighter and bomber were born. As wartime demands became greater, aircraft designs advanced rapidly. Airplanes became larger and faster, and many pilots gained fame through their exploits in aerial combat as both sides attempted to gain control of the sky.

During World War I, Hazelhurst Field (formerly Hempstead Plains) and neighboring Mitchel Field on Long Island were two of the Army Air Service's largest and most-important airfields. Hundreds of aviators were trained for war at these two fields, and important aeronautical experimentation took place as well. During the war, many new hangars and other buildings were constructed, and the flying fields were expanded and leveled into the general shape they would retain through the 1950s.

HANGAR ROW ON HAZELHURST FIELD, C. 1918. The aircraft are Curtiss JN-4 Jennies. The World War I airfields were considerably larger than the prewar ones, although they retained the same plan of simple wooden buildings and hangars and a large grassy flying area.

AN AVIATOR WITH HIS CURTISS JN-4 JENNY AT MITCHEL FIELD, C. 1918. The JN-4 was the most widely used American training aircraft during World War I, and almost 4,000 of them were built. It was powered by a 90-horsepower Curtiss V-8 engine and cruised at about 50 miles per hour. Jennies were the most common aircraft in Long Island skies during this period.

AN INVERTED CURTISS JN-4 ON MITCHEL FIELD, 1918. Crashes such as this one were usually due to the plane flipping over upon landing, probably because it hit a rut or soft spot. The pilot usually walked away.

A Curtiss JN-4 Crash in a Long Island Field, c. 1918. The fact that this crash appears to have happened in a freshly plowed field usually means that the aircraft probably suffered an engine failure.

A Fatal Crash of Privates Meritt and Spilena at Hazelhurst Field, May 7, 1917.
This crash was caused by two men who apparently were not pilots taking up one of the training machines. After a steep dive, they pulled out too abruptly, causing the aircraft to suffer an in-flight structural failure.

A Wrecked JN-4, 1918. This plane was deposited in front of the maintenance hangar at Mitchel Field.

A CURTISS JN-4, C. 1918. This wreck occurred while attempting a forced landing on a Long Island road.

A CURTISS JN-4 UPENDED ON HAZELHURST FIELD, C. 1918. The aircraft appears to have hit a muddy spot in the field while landing.

A Curtiss JN-4 Mail Plane after a Forced Landing in Queens, c. 1918. Immediately after World War I, the Curtiss JN-4 became the first plane of the U.S. airmail service. The New York terminus for the airmail was at the Belmont Park Racetrack in Elmont, Long Island.

A Curtiss JN-4 Mail Plane Wrapped around a Tree, Nassau County, c. 1918. This accident was also probably caused by an engine failure.

An Airmail Plane Crash, 1918. Pilots and a ground crew gather after a nonfatal airmail plane crash, still burning in the rear, at Belmont Park.

A DeHavilland DH-4 over Mitchel Field, c. 1920. Between 1918 and 1925, the most common aircraft on U.S. Army aviation fields was the DH-4. Although a British design, almost 5,000 were built in the United States in 1917 and 1918. Powered by a 420-horsepower Liberty engine, the aircraft cruised at about 90 miles per hour. They were used for training, scouting, and bombing.

A Fatal DH-4 Crash on Mitchel Field, c. 1918. The aircraft appears to have spun in from altitude.

DH-4 CRASHES, 1918. The above DH-4 crashed while attempting a forced landing in a Long Island hayfield. The DH-4 seen below crashed while trying to land on a frozen pond. Such accidents were usually caused by engine failure.

A DH-4 THAT CRASHED AND BURNED, C. 1918. Because of its flammability and the location of its fuel tank, the DH-4 was also known as "the flaming coffin."

UPENDED IN A MUDDY SPOT ON HAZELHURST FIELD, C. 1918. Upon the war's end, DH-4s were pressed into service as mail planes. Hazelhurst Field became the central point for converting military DH-4s into mail planes. Aircraft nosing over in muddy spots was a common occurrence on the dirt and grass fields of the day.

A BRITISH HANDLEY PAGE 0400 BOMBER ON HAZELHURST FIELD, 1918. During World War I, Hazelhurst Field was the main U.S. Army Air Service experimentation center. Virtually any new piece of aviation equipment and many aircraft were tested here. As the army was looking to place a heavy bomber into service, and none were in production in the United States, several foreign types were tested.

THE CRASH OF A HANDLEY PAGE, HAZELHURST FIELD, 1918. Because of the lack of familiarity with large aircraft, operating these foreign types proved difficult, as the wreck of this bomber shows.

AN ITALIAN CAPRONI CA-36 BOMBER ON HAZELHURST FIELD, 1918. Another promising heavy bomber candidate was the Caproni CA-36. This type saw widespread and successful use on the Italian front, so one was tested at Hazelhurst Field in 1917 and 1918.

THE CRASH OF THE CAPRONI CA-36, HAZELHURST FIELD, 1918. The Caproni crashed on May 17, 1918, killing its pilot, Captain Resnati. The reasons for the crash are unknown. By the end of 1918, the air service decided to develop an American aircraft for heavy bomber use.

Three

THE GOLDEN AGE
1919–1939

During the 1920s and 1930s, the technology of aircraft underwent a revolution. Flying went from being a dangerous sport to becoming a viable major industry. Aircraft themselves evolved from being constructed of wood and fabric to steel and aluminum. During the 1920s, aviation began to touch all aspects of American life. The public clearly saw the tremendous potential of the airplane for commercial transport, airmail, cartography, and sport. All of these trends manifested themselves on Long Island. Approximately 20 Long Island aircraft manufacturers made major contributions to national civil and military aviation.

During this period, Long Island also had the most airfields of any geographic area in America. They were among the busiest. Mitchel Field remained the army's premier facility, while nearby Roosevelt Field was the most active civil airfield in the country. In the 1930s, Roosevelt Field, then known as the "world's premier airport," became the most technically advanced civil airfield in the United States. Over 450 planes were based there, and virtually every civil aircraft manufacturer had a distributor there. It became the scene of numerous historic flights and record-setting events.

LOOKING WEST ACROSS MITCHEL FIELD, C. 1925. Through the 1920s, Mitchel Field still retained its World War I structures and grass flying area. Right through this period, it remained one of the army's largest and most active fields.

AN AERIAL VIEW OF MITCHEL FIELD, 1939. In the early 1930s, Mitchel underwent a huge renovation and expansion program. It now had many fine steel and concrete hangars, new brick housing, maintenance and administrative buildings, and long concrete runways.

THE ONE AND ONLY L.W.F. OWL BOMBER AT MITCHEL FIELD, 1920. Built in nearby College Point, this was then the largest land plane built in America to date. Despite its size and power, it displayed poor flight characteristics.

THE CRASH OF THE L.W.F. OWL AT MITCHEL FIELD, 1921. After nosing over in a rut, the Owl sustained some fairly severe damage. However, it was repaired and flew until 1923.

33

THE CURTISS NBS-1 BOMBER CRASH AT MITCHEL FIELD, 1923. The NBS-1 was a Curtiss-built version of the famous Martin MB-2 bomber of 1919. They remained as front-line bombers in the army inventory until 1928.

A FATAL DH-4 CRASH, MITCHEL FIELD, 1924. Through the 1920s, DeHavilland DH-4s had a major presence at Mitchel. They were used primarily for observation and bomber-training duties.

A DH-4 CRASH, 1923. Apparently, the plane spun in at a fairly high speed in this fatal crash.

LIEUTENANT GRIFFISS'S DH-4 CRASH, MITCHEL FIELD, 1923. Such accidents were usually caused by the plane hitting a soft spot in the field while landing.

HITTING THE HANGAR, 1923. Seen is a DH-4 after striking the side of a hangar at Mitchel.

THE CRASH OF LIEUTENANT STRICKLAND'S THOMAS MORSE MB3A, 1924. This crash occurred at Mitchel Field. The MB3A was one of the first American post–World War I fighters. Thomas Morse and Boeing built 250. They served between 1919 and 1928.

A C-9 TRANSPORT UP ON ITS NOSE AT MITCHEL, 1930. The C-9 was the military transport version of the famous Ford Trimotor.

A Curtiss 0-1 Falcon at Mitchel Field, c. 1928. The two-place Falcon was the most predominant type of aircraft based at Mitchel Field in the late 1920s. They were primarily used for observation and photo-reconnaissance duties. Due to their number on the field, most of the Mitchel aircraft accidents during this period involved Falcons.

THE CURTISS FALCON, C. 1928. These Falcon crashes occurred on and near Mitchel Field.

A CURTISS FALCON CRASH, 1929. This Falcon crashed on the roof of a Westbury house.

LT. TIMOTHY CREEDON AND CPL. GILBERT BURNETTE. In Port Washington, on May 14, 1928, Creedon and Burnette were involved in the fatal crash seen here. The pilot became lost in a heavy rainstorm while trying to find Mitchel Field. He attempted a forced landing at night on the beach of a north shore estate. Unfortunately, the plane struck a tree and buried its nose in the soft ground.

A MARTIN B-10 BOMBER AT MITCHEL FIELD, C. 1935. The B-10 was the army's premier bomber in the late 1930s, and at least one squadron of them was consistently stationed at Mitchel Field during this period. Their primary mission was defending the coast around the U.S. mainland.

THE CRASH OF A MITCHEL FIELD MARTIN B-10 IN SYOSSET. On June 6, 1936, this aircraft went into a spin. The crew bailed out safely.

A DOUGLAS O-35 CRASH NEAR MITCHEL FIELD, JULY 13, 1933. This nonfatal crash involved an O-35, one of only 15 of these armed reconnaissance aircraft built.

A KEYSTONE B-4 BOMBER CRASH, 1931. These aircraft were considered light bombers, with about 100 produced for the U.S. Army Air Service between 1928 and 1932. One squadron of B-4s was stationed at Mitchel Field.

A VOUGHT VE-7 SCOUT PLANE, C. 1923. This plane flipped over in the surf just off Rockaway Naval Air Station.

A SEVERSKY BT-8 TRAINER. This aircraft ground-looped and flipped over at the company's East Farmingdale plant in 1936. The BT-8 was the company's first mass-produced aircraft. It ultimately led to the P-35 and later the famous P-47 fighter.

A PROTOTYPE GRUMMAN F3F FIGHTER, 1935. This photograph was taken at the Grumman Farmingdale plant. About 170 of these fighters were built for the U.S. Navy through 1939. These were probably the greatest biplane fighters ever built.

THE FATAL CRASH OF THE PROTOTYPE GRUMMAN F3F. On March 22, 1935, in Farmingdale, Grumman test pilot Jimmy Collins was conducting a high-speed dive test when the aircraft suffered structural failure.

A VIEW OF ROOSEVELT FIELD, LOOKING NORTH, C. 1928. From the late 1920s through the 1930s, this was by far the busiest general aviation airfield in the United States. It had the most civilian aircraft based here and the most operations per hour. There were, however, no air-traffic control and no paved runways.

ROOSEVELT FIELD, C. 1937. At the time of this photograph, the field had been upgraded with many new buildings, services, concrete hangars, paved runways, airfield lighting, and rudimentary air-traffic control.

A JUNKERS LARSEN JL-6 AT MITCHEL FIELD, C. 1922. An early all-metal design built in Germany as the F-13 airliner, the Junkers was imported into the United States by John Larsen, who set up a distributorship at Central Page (Bethpage) Flying Field. Larsen imported 25 of these planes between 1920 and 1922, most of which were sold to the postal service for use as mail planes.

THE WRECK OF A JL-6 NEAR CENTRAL PARK FLYING FIELD, 1922. Several of the JL-6 aircraft crashed due to faulty fuel lines that ruptured, which caused engine fires.

A SIKORSKY S-35 TRANSATLANTIC AIRCRAFT, ROOSEVELT FIELD, 1926. Roosevelt Field became the home to many aviators who attempted to span the Atlantic in the late 1920s. This aircraft was built by Igor Sikorsky for a transatlantic flight attempt by a World War I ace pilot, Rene Fonck.

THE REMAINS OF THE SIKORSKY S-35, SEPTEMBER 21, 1926. While attempting a takeoff from Roosevelt Field with a full load of fuel for a nonstop flight to Paris, the landing gear of the S-35 collapsed, causing it to cartwheel and burn, killing two of the crewmembers on board.

Starting the Sellars Quadroplane, July 20, 1927. Built by Dr. J. Sellars, this was a tiny quadroplane design made entirely of wood and powered by an eight-horsepower engine. It weighed but 150 pounds. This photograph was taken at Roosevelt Field.

The Sellars Quadroplane, July 20, 1927. While attempting to start the plane's engine, gasoline ran on to the cockpit floor, where a backfire ignited it. The pilot survived with minor burns.

VIOLA GENTRY'S PARAMOUNT CABINAIRE, THE ANSWER, 1929. This photograph was taken at Roosevelt Field. In this aircraft, Viola Gentry, along with Jack Ashcraft, hoped to set a new world refueled endurance record.

THE WRECK OF THE ANSWER IN WESTBURY, JUNE 28, 1929. Less than 10 hours after takeoff, the *Answer* ran out of fuel at night, as a fog had prevented the refueling plane from taking off. The plane struck a tree at Hicks Nursery while attempting a night landing in zero visibility. Ashcraft was killed and Gentry was seriously injured.

THE GULL AT ROOSEVELT FIELD, 1928. This aircraft, designed by Leonard Bonney, was a radical design with folding wings whose incidence could also be adjusted in flight. He based his design on the seagull and reportedly had it built at a cost of $100,000.

THE REMAINS OF THE GULL, ROOSEVELT FIELD, MAY 4, 1928. Attempting his first flight in the *Gull*, Bonney rose about 50 feet, whereupon the aircraft immediately dove into the ground. Bonney was killed and the aircraft was destroyed. Here, the wreckage has been dumped behind a hangar on the field. Someone is apparently looking through it for usable parts.

51

April 27, 1924

THE AMERICAN EAGLE CRASH, NEAR ROOSEVELT FIELD, APRIL 27, 1929. Two men were killed when this aircraft struck a tree, crashed, and burned.

A Fokker F-32 Airliner, 1929. Built in New Jersey by Fokker, this 100-foot-long aircraft was large for its day and was powered by two tractor and two pusher engines. A total of seven were built, and they received a generally unenthusiastic response from the airlines.

The Crash of the Fokker F-32, Carle Place, November 27, 1929. While demonstrating a three-engined takeoff from Roosevelt Field, one of the plane's port engines was stopped. The other port engine failed, and the aircraft crashed into a suburban neighborhood in nearby Carle Place. Remarkably, no one was killed, but the pilot and a passenger were injured.

A CURTISS ROBIN AT ROOSEVELT FIELD, 1928. One of the most successful civil aircraft designs of the late 1920s was the Curtiss Robin, designed and first built in Garden City. The Robin was a forerunner of modern civil aircraft in that it was a monoplane with an enclosed cabin. Altogether, 769 Robins were built, making them the most numerous civil aircraft of that time period.

A CURTISS ROBIN CRASH, ROOSEVELT FIELD, 1932. Note the muddy and icy flying field. The plane probably crashed due to a ground loop.

54

THE RESULT OF A COLLISION WHILE LANDING, ROOSEVELT FIELD, C. 1935. Due to the total lack of air-traffic control, collisions on or near the field were not uncommon. The aircraft in the foreground is a Taperwing Waco.

A BARREN ISLAND CRASH, C. 1930. Seen in this photograph is a fatal aircraft crash at Barren Island, Brooklyn.

A BURNING AIRCRAFT, ROOSEVELT FIELD, C. 1930. This monoplane probably nosed over on the snowy flying field, its fuel spilling out and starting an engine fire.

A FATAL AIRCRAFT CRASH, WESTBURY, C. 1930. This biplane burned upon crashing. Two bodies are covered over with sheets.

A Brunner Winkle Bird at Roosevelt Field, 1928. Built in Queens, the Bird was a fairly successful design able to carry three people. The early versions used World War I surplus OX-5 engines; the later models used K-5 Kinners. Before the firm was forced to close due to the Depression, 220 of these planes were built.

BRUNNER WINKLE BIRD ACCIDENTS, C. 1930. These crashes, involving Brunner Winkle Birds, took place on and near Roosevelt Field.

A Bird Crash, June 7, 1931. When this Bird crashed into a house on Remsen Street, Hempstead, due to engine failure, both of the home's occupants escaped without injury.

A Beech Staggerwing, c. 1940. This aircraft is seen nosed over on Roosevelt Field.

A FATAL SKYDIVING ACCIDENT, NEAR ROOSEVELT FIELD, 1936. Along with being a major civil airport, Roosevelt Field was also the original home of the U.S. Parachute Association. Thus, skydivers were also a common sight over Roosevelt Field. In the 1930s, skydiving was a much more dangerous sport than it is today.

A PRIMARY GLIDER AT ROOSEVELT FIELD, C. 1928. The cheapest way to get into the air in the late 1920s and early 1930s was in a simple primary glider. These planes were usually built at home from purchased plans. They were towed into the air behind an automobile or catapulted into the air with bungee cords. The pilots usually had no previous flying experience.

A Primary Glider Crash, 1929. This crash, at Fitzmaurice Field in Massapequa, was fatal.

A Primary Glider Crash in Queens. This primary glider crashed into the roof of a home in Queens on August 4, 1934.

A PIPER J-3 CUB AT ROOSEVELT FIELD, C. 1939. Probably the most famous and most popular civil light plane ever built was the Piper Cub. Production began in 1937, and by the time it ceased in 1947, more than 20,000 Cubs had been built. The plane could carry two and was powered by a 65-horsepower engine.

A FATAL PIPER CUB CRASH, WESTBURY, C. 1947. Bordering the northern edge of Roosevelt Field, Westbury was a frequent site of air crashes between 1930 and 1950.

A FATAL PIPER CUB CRASH, EAST MEADOW, 1940. Elliot DeLister was practicing spins when he lost control of the aircraft and spun into an East Meadow house.

A PIPER CUB CRASH, DEER PARK, C. 1950. This aircraft was based at Deer Park Airport, which was one of Long Island's larger general aviation airports in the 1950s and 1960s. It closed in 1974.

A Piper Cub Crash, Bayshore, c. 1940. This Cub wound up in the house's open garage.

A Travelaire E-4000 up on Its Nose at Roosevelt Field, c. 1935. Such accidents were almost always caused by hitting a soft spot in the field.

THE CRASH OF A STINSON RELIANT, ROOSEVELT FIELD, C. 1938. The Reliant, seating four, was one of the larger, heavier, and more expensive general aviation aircraft available in the 1930s.

A CURTISS FLEDGLING AT ROOSEVELT FIELD, 1929. Derived from the navy's N2C-1 trainer of 1928, the Fledgling was built as a primary trainer for the Curtiss Flying School chain. The largest of the Curtiss flying schools was on Roosevelt Field. A total of 110 Fledglings were built.

A Curtiss Fledgling after a Fire, 1929. The aircraft appears to have made a forced landing on the golf course just east of Mitchel Field. After the landing, the fabric-covered plane burned up.

CURTISS FLEDGLING CRASHES, ROOSEVELT FIELD, C. 1930. Fledglings were rarely seen outside of the Curtiss Flying School chain. They were designed to replace the aging Curtiss Jennies in the 1920s. However, students found the aircraft too large, heavy, and difficult for basic flight training.

A CABIN WACO UIC AT ROOSEVELT FIELD, C. 1933. This was a four-place enclosed-cabin Waco that was known as being rugged and dependable. Quite a few were based at Roosevelt Field in the 1930s.

A CABIN WACO UP ON ITS NOSE, ROOSEVELT FIELD, C. 1935. Standing an aircraft up on its nose was a problem that plagued tail wheel aircraft, especially on the older dirt fields. Modern general aviation aircraft have a nose wheel that has eliminated the problem.

A TP SWALLOW CRASH AT ROOSEVELT FIELD, C. 1929. This aircraft was owned by noted aviatrix Viola Gentry, who can be seen kneeling by the engine.

THE AERONCA C-3 AT ROOSEVELT FIELD, C. 1934. The C-3 was a popular two-place light plane powered by a two-cylinder Aeronca engine. About 500 were built between 1932 and 1935. They were used primarily for flight training and club flying.

AN AERONCA C-3 ON ITS BACK AT ROOSEVELT FIELD, C. 1935. The C-3 was also known as "the Flying Bathtub," as the two-place cockpit was slung low, just above the axle. This made the aircraft harder to flip, but not impossible.

70

A FAIRCHILD 22, C. 1933. A two-place parasol monoplane, the Fairchild 22 enjoyed some popularity as a sport plane. Between 1931 and 1934, 125 were built in several versions.

A FAIRCHILD 22 CRASH AT ROOSEVELT FIELD, C. 1935. This aircraft has flipped on its nose and then slammed back down. Undoubtedly, the occupants were seriously injured.

A Fairchild 22 Crash, Westbury, June 24, 1938. In this crash, Capt. Harry Manning, former navigator for Amelia Earhart, was seriously injured.

A Fatal crash of a Taylor Cub, near Roosevelt Field, April 8, 1936. John Mendelo was killed when his Cub collided with a Bird biplane flown by Walter Barlow in the traffic pattern near Roosevelt Field.

THE CRASH OF A BRUNNER WINKLE BIRD, NEAR ROOSEVELT FIELD, APRIL 8, 1936. This Bird, flown by Walter Barlow, collided with the Cub seen in the previous photograph, at a 500-foot altitude near Roosevelt Field. Barlow was seriously injured.

A GOODYEAR NONRIGID TYPE PA AIRSHIP, C. 1932. In the early 1930s, Goodyear built six similar airships, with 112,000 cubic feet of helium, for advertising and publicity purposes.

THE CRASH OF GOODYEAR TYPE PA AIRSHIP COLUMBIA. This airship did aerial advertising and the earliest live aerial traffic reports while based at Holmes Airport. The airship was less than a year old when this crash took place, near Holmes Airport in Queens on February 13, 1932. A mechanic was killed.

A DeHavilland DH-60 Gypsy Moth, Roosevelt Field, c. 1932. Beginning in the mid-1920s, the DH-60 was the first personal plane to be mass-produced. Some 1,760 were built in England between 1925 and 1932, with an additional 150 built under license by Curtiss Wright in the United States.

A DH-60 Gypsy Moth Crash, 1931. This crash, which was fatal, took place on Jones Beach.

FLEET MODEL 2, ROOSEVELT FIELD, C. 1930. The Fleet 2, powered by a Kinner engine, was a popular training plane in the late 1920s and early 1930s. The Roosevelt Aviation School at Roosevelt Field purchased five of them for use as basic trainers. For winter operations, they were fitted with skis, as seen here. Due to the popularity of the Fleet on Long Island, as elsewhere, they were involved in a number of crashes.

A ROOSEVELT AVIATION SCHOOL FLEET ON ITS BACK, ROOSEVELT FIELD, C. 1930. The two problems that plagued the Fleet were an engine prone to failure and a lack of directional stability, which often led to an accidental spin when making a steep turn.

A ROOSEVELT AVIATION SCHOOL FLEET. This aircraft was involved in a nonfatal forced landing in East Meadow in 1932. This was the school's night flying trainer, as can be seen by the landing light hung up in the tree.

A FATAL FLEET CRASH, FREEPORT, 1931. The aircraft appears to have spun in, vertically striking a house on the Great South Bay.

THE FORCED LANDING OF A FLEET NEAR MASSAPEQUA PARK, 1931. After losing power, this Fleet, piloted by Paul Saltanis, attempted a landing in a small field and wound up in the trees at the far end. The two people on board suffered minor injuries.

A FLEET CRASH ON POST AVENUE, WESTBURY, NOVEMBER 1930. Not a forced landing, this aircraft appears to have hit the middle of the street while in a spin, not an uncommon occurrence in a Fleet.

A FLEET FORCED LANDING, CARLE PLACE, 1932. The aircraft has obviously struck a pole or a tree while making a forced landing.

A CESSNA AW CRASH, BELLMORE, 1931. Noted black aviator and activist Sufi Hamid and a passenger were killed in this crash.

A MONOCOUPE 90 AT ROOSEVELT FIELD, C. 1935. Originally designed as a racing plane, the two-place Monocoupe was fast and nicely streamlined. Over 150 were built between 1932 and 1935.

A FATAL MONOCOUPE CRASH NEAR ROOSEVELT FIELD, C. 1935. Powered by a Lambert engine, Monocoupes were generally safe aircraft, and they dominated the racing scene in the early 1930s. They were clearly rich men's airplanes.

A KELLETT K-2 AUTOGIRO AT ROOSEVELT FIELD, C. 1935. The autogiro was a direct ancestor of the helicopter. An aircraft engine and propeller pull the machine forward while unpowered freewheeling rotors lift it up and down at steep angles.

A KELLETT AUTOGIRO CRASH NEAR ROOSEVELT FIELD, C. 1935. As the autogiro was a new kind of flying machine, inexperienced pilots often got into trouble with them. A crash by Amelia Earhart in one severely hurt their popularity.

A Stinson SR Inverted at Roosevelt Field, c. 1935. The Stinson SR Reliant was one of the first general aviation aircraft to feature landing flaps. However, pilots had difficulty getting used to the new devices, which required a much steeper landing angle.

A Stinson 105 Voyager Crash, near Roosevelt Field, 1941. It appears to have been a successful forced landing in a small depressed area.

A FATAL WACO 10 CRASH, WESTBURY, C. 1930. The Waco 10 was a fairly successful three-place biplane powered by a World War I surplus OX-5 engine. Over 1,100 were sold between 1927 and 1929.

A TAYLOR E-2 CUB, ROOSEVELT FIELD, C. 1934. A fairly successful civil light plane, 350 E-2 Cubs were built between 1931 and 1935. They were powered by 40-horsepower Continental engines.

REMAINS OF A TAYLOR E-2 CUB, 1935. Here, remains of an E-2 Cub are in a tree near Roosevelt Field after a midair collision.

A PIPER CUB FLOATPLANE AFTER A CRASH, WANTAGH, 1940. As Long Island is surrounded by water, a great number of floatplanes operated here in the 1930s. Floatplanes suffered the same proportion of accidents as did land planes.

A Piper Cub Floatplane Crash near Long Beach, c. 1940. Floatplane crashes usually occurred during landing when pilots misjudged their height above glassy water surfaces.

A Wrecked Taylor Cub on a Dade Truck, Roosevelt Field, c. 1938. With so many air crashes on Long Island, one wonders what became of the planes. In the 1930s, a local trucking and construction firm, Dade Brothers, developed a lively business removing wrecked aircraft. The planes were usually taken to Roosevelt Field, salvaged for any usable parts, and then unceremoniously burned behind one of the hangars.

A WRECKED LOCKHEED ELECTRA ON A DADE TRUCK, ROOSEVELT FIELD, C. 1938. The Electra was designed as a small twin-engine airliner and was produced in several versions between 1934 and 1941.

A DADE TRUCK REMOVES A DC-3. In this 1940 photograph, a wrecked American Airlines DC-3 is loaded on a Dade truck at LaGuardia Airport.

Four

WORLD WAR II
1939–1945

During World War II, two Long Island aircraft manufacturers, Grumman and Republic, built fighters that were among the finest ever produced. The aircraft rolled off Long Island assembly lines in record numbers and clearly helped America and its allies achieve victory in the war. On the home front, Mitchel Field continued to remain one of the most important army airfields, and P-40s and B-25s stationed here provided the air defense for New York City and America's northeast coast.

Local newspaper reports reveal that there were numerous military aircraft crashes on Long Island during the war, both by Grumman and Republic aircraft being tested and patrol plane accidents out of Mitchel Field. However, due to wartime censorship, photographs of such accidents are extremely rare. The few examples presented here are but a small number of those that actually occurred, but they typify wartime aircraft accidents.

THE LOCKHEED P-38 LIGHTNING. The Lightning was an Air Corps fighter produced from 1939 to 1945. A radical design for its day, the Lightning served in every combat theater and was able to achieve speeds of 400 miles per hour.

On February 11, 1939, the one and only XP-38, flown by Lt. Ben Kelsey, was on a record-setting, seven-hour cross-country flight when it suffered carburetor icing on approach to Mitchel Field and crashed just short of the runway.

A DOUGLAS B-18 BOMBER AT MITCHEL FIELD, C. 1939. In production from 1936 to 1939, B-18s were based at Mitchel Field through 1941 as bomber and observation aircraft. They were derived from the civilian DC-2.

A MIDAIR COLLISION OF TWO B-18S OVER QUEENS, JUNE 17, 1940. On this day, two B-18s out of Mitchel Field on a training mission collided over 239th Street and 87th Avenue in Bellerose, Queens, killing 11 people in the aircrew and a person on the ground.

CURTISS P-40B WARHAWKS ON MITCHEL FIELD, 1941. In production between 1938 and 1944, almost 10,000 P-40s were built during World War II. They were stationed at Mitchel Field throughout the war to provide the air defense for New York City. The majority of aircraft crashes on Long Island between 1940 and 1945 were Curtiss P-40s.

A P-40 CRASH IN EAST FARMINGDALE, DECEMBER 8, 1941. In this crash, the pilot was killed while attempting a forced landing after suffering an engine failure. He struck a car while trying to land in an open field.

A REPUBLIC P-43 LANCER FORCED LANDING, C. 1941. This aircraft's forced landing took place somewhere on Long Island. A predecessor to the famed P-47 Thunderbolt, 372 P-43s were built on Long Island between 1940 and 1941. None saw combat in American service during the war.

A REPUBLIC P-47D THUNDERBOLT AT THE PLANT, C. 1943. With over 15,000 built, the P-47 was the most-produced American fighter of all time. It was also the largest and heaviest fighter of World War II. It was in record-setting production between 1941 and 1945. This photograph was taken at the Republic plant in East Farmingdale.

THE RESULTS OF AN ENGINE FAILURE, C. 1943. This crash occurred over Long Island while testing a new P-47D. The aircraft is probably being shipped back to the factory.

A P-47D CRASH DURING A FLIGHT TEST, C. 1944. The aircraft has arrived back at the factory to see if anything can be salvaged.

A NORTH AMERICAN B-25 MITCHELL, C. 1941. In production between 1940 and 1945, the B-25 was the standard Air Corps medium bomber during World War II. A total of almost 11,000 were built. Primarily for antisubmarine patrol, B-25s were stationed at Mitchel Field from late 1941 through the end of the war.

A B-25 CRASH, EAST MEADOW, SEPTEMBER 2, 1943. After returning from tow target duty, this B-25 suffered engine failure and crashed in an open field about a mile east of Mitchel Field, killing its crew of three.

A Grumman F4F Wildcat Crash, c. 1946. The F4F Wildcat was the navy's front-line fighter in the first years of World War II. This is an FM-2 version built under license by General Motors. The aircraft is probably privately owned, purchased as surplus right after the war, and has made a wheels-up landing somewhere on Long Island.

A Grumman F6F Hellcat Crash, c. 1946. Grumman's F6F was the best naval fighter of World War II and one of the greatest of all time. This one, based at the naval air station at Floyd Bennett Field in Brooklyn, has made a forced landing in the surf off Long Beach.

Five

THE POSTWAR YEARS
1946–1959

The end of World War II saw explosive growth in commercial aviation on Long Island, while many civil and military airfields were forced to close due to rapid population growth and rising land values. LaGuardia and Idlewild (now Kennedy) Airports became the busiest commercial airfields in America, while the long-established Roosevelt and Mitchel Fields ceased operation due to financial and political pressure. Several spectacular military air crashes ultimately did in Mitchel Field, while civil aircraft accidents became increasingly rarer due to the declining number of small aircraft and steadily increasing air safety. However, for the first time, commercial aircraft accidents now became an occasional occurrence due to the rapid growth of the airline industry. Local aircraft manufacturing continued at a steady pace during this period, so accidents while testing new military aircraft continued to occur with some rarity. In recent years, general aviation accidents have become extremely rare due to good air-traffic control, new instrumentation, reliable engines, and safer aircraft designs.

AN XR-12 RAINBOW LANDING ACCIDENT. This accident occurred at Republic Airport on July 9, 1947. The XR-12 was a U.S. Air Force experimental high-altitude photo-reconnaissance aircraft. On this test flight, one of its main landing gears failed to lower.

A SIKORSKY HOS-1 CRASH, BROOKLYN, MAY 9, 1947. The first helicopter training station for the U.S. Navy was at Floyd Bennett Field in Brooklyn. This Sikorsky, flown by Lt. David Gershowitz, crashed during a demonstration flight. Its pilot survived.

A Grumman F8F Bearcat, c. 1947. This was the best piston-engined fighter to see service with the U.S. Navy. It was in production by Grumman on Long Island from 1944 to 1949.

A Fatal Crash of a Grumman F8F, Hicksville, April 14, 1947. This aircraft crashed on Willis Lane in Hicksville while on a test flight. Other than the pilot, no one was hurt.

A REPUBLIC P-84 THUNDERJET, C. 1947. The P-84 was an early postwar American jet fighter and bomber. Over 7,500 were built between 1945 and 1953 in Farmingdale.

A THUNDERJET LANDING ACCIDENT, MARCH 1947. This YP-84A, flown by Carl Bellinger, had an accident at Republic Airport in Farmingdale when the brakes failed upon landing. The aircraft went through a fence, struck several vehicles, and flipped over. Bellinger was lucky, as the cockpit came down between two parked cars. He walked away.

A P-84 ENGINE FIRE, FARMINGDALE, 1947. This Thunderjet was being run up on Republic Field when an engine fire erupted due to a broken fuel line. There were no injuries.

AN AERIAL VIEW OF MITCHEL FIELD, LOOKING SOUTHWEST, C. 1955. After World War II, Mitchel Field remained an important airbase as the headquarters of the Continental Air Command. However, the open expanse that surrounded it when it was originally built was now replaced by a sea of houses. There was no margin for error.

A FATAL B-25 CRASH, SEPTEMBER 13, 1955. This crash took place at Greenfield Cemetery in Hempstead. Six people were killed when the aircraft suffered an engine failure on takeoff from Mitchel.

A B-25 FORCED LANDING, EAST MEADOW, MAY 25, 1953. Another crash is caused by engine failure on takeoff from Mitchel. This time, the crew of three walked away.

NORTH AMERICAN F-82 TWIN MUSTANGS, MITCHEL FIELD, C. 1950. These aircraft were stationed at Mitchel in the late 1940s and early 1950s. They were developed as a very long-range fighter at the end of World War II. Only 250 were built through 1947, and the type saw some action in Korea.

AN F-82 CRASH, HEMPSTEAD, MAY 4, 1949. This F-82 crashed when Lt. Andrew Wallace suffered an engine failure on takeoff. The aircraft wound up on the front lawn of a house on Duncan Road in Hempstead. Both the pilot and radar operator survived.

AN F-82 CRASH, UNIONDALE, OCTOBER 1949. After engine failure on takeoff, this aircraft wound up on Hempstead Turnpike. The pilot escaped before the aircraft burst into flames.

A DOUGLAS A-26 INVADER OVER LONG ISLAND, C. 1950. The A-26 was in production from 1942 through 1945, and it was the fastest attack aircraft of World War II. The type remained in the U.S. Air Force inventory for close air support through 1965.

AN A-26 CRASH, UNIONDALE, SEPTEMBER 14, 1955. After losing an engine on takeoff, this A-26 wound up in the middle of Hempstead Turnpike. Luckily, there were no injuries.

A FATAL A-26 CRASH, EAST MEADOW, NOVEMBER 3, 1955. After losing power, this A-26 crashed on Barbara Drive in East Meadow, killing its crew of two. Fortunately, there were no injuries on the ground.

A BEECHCRAFT C-45 FORCED LANDING, MARCH 13, 1949. This forced landing was just short of the Mitchel Field runway. The C-45 was used as a personnel transport plane by the U.S. Air Force from the late 1940s into the 1960s.

AN F-86 CRASH, LEVITTOWN, DECEMBER 27, 1955. This Sabre Jet exploded and crashed on Blacksmith Road in Levittown. The pilot ejected safely, and by a miracle, there were no casualties on the ground.

A FATAL T-33 CRASH, UNIONDALE, OCTOBER 6, 1956. This T-33 landing at Mitchel overshot the runway, hit the perimeter fence, and flipped over on Hempstead Turnpike. The pilot, Maj. Daniel Kramer, was killed, and three people in a car were injured.

A C-124 LANDING ACCIDENT, MITCHEL FIELD, FEBRUARY 1, 1957. The Douglas C-124 Globemaster was the U.S. Air Force's main personnel and cargo aircraft of the 1950s.

A FAIRCHILD C-123 PROVIDER, C. 1960. This aircraft was an air force troop and cargo carrier from the mid-1950s through the 1960s. The type saw widespread use in Vietnam.

A FATAL C-123 CRASH, LINDENHURST, OCTOBER 16, 1958. This was one of Mitchel Field's most notorious air crashes. A C-123 flying from Bradley Field, Connecticut, to Mitchel Field ran out of fuel a few miles short of the field. The aircraft attempted a forced landing on the Southern State Parkway in Lindenhurst, where it struck three cars and went through an underpass, ripping the wings off. One motorist was killed, and four others were injured.

A PIPER CUB FLOATPLANE CRASH, ISLAND PARK, 1947. General aviation accidents diminished noticeably during this period, but they still occurred on a fairly regular basis. Here, a J-3 Cub has flipped over on landing in Reynolds Channel, Island Park.

A COLLISION ON THE GROUND BETWEEN TWO PIPER CUBS, FLUSHING AIRPORT, 1949. Flushing Airport was the last general aviation airport within the limits of the city of New York. It closed in 1984 mainly due to its proximity to LaGuardia Airport.

A Fatal Piper Cub Crash, Oceanside, 1951. Although the Piper Cub was first produced in the late 1930s, it was one of the most popular light planes ever built. Thousands are still flying.

A Nonfatal Stinson Voyager Crash, Massapequa Park, 1954. The Stinson Voyager was one of the few unspinnable aircraft ever built, so accidents such as this were usually caused by engine failure.

A Piper Cruiser Crash, Long Island City, December 19, 1959. Here, a Cruiser flown by aviation photographer Anthony Hanzlik suffered an engine failure but made a remarkable forced landing in Roosevelt Park, Long Island City. Hanzlik escaped with minor injuries.

A Nonfatal Stinson Crash, Roosevelt Park, Oyster Bay, c. 1960. Roosevelt Park, on the southern edge of Oyster Bay, is not that large and is generally full of trees. Thus, this forced landing was fairly well executed.

LaGuardia Airport, Queens, 1951. In the late 1940s and early 1950s, this was the busiest commercial airport in the world. At this same time, nearby Idlewild (now Kennedy) Airport was developing into one of the largest international airports in the world.

AN AMERICAN AIRLINES DC-3, FEBRUARY 1947. Seen is an American Airlines DC-3 after making a forced landing on Jones Beach.

A FATAL DC-4 CRASH, LAGUARDIA AIRPORT, MAY 30, 1947. A United Airlines DC-4 attempted to take off with the control locks still on. The plane failed to gain altitude and crashed, killing 39 people.

A LOCKHEED CONSTELLATION FORCED LANDING, MITCHEL FIELD, AUGUST 30, 1949. A Pan American Lockheed Constellation from New York (Idlewild) on its way to London was involved in a midair collision with a Cessna over central Long Island. The Constellation made an emergency landing at Mitchel Field, where it was repaired and flown out six weeks later. The Cessna crashed into the galley area, so there were no fatalities on the Constellation. However, the two occupants of the light plane were killed.

A Douglas DC-6, c. 1947. During the first 25 years of the commercial jet age, the DC-6 was the most sought-after piston-engined transport. The type was in production between 1946 and 1951, and the planes were widely used for both passenger transport and freight operations.

A DC-6 FORCED LANDING, JAMAICA BAY, JUNE 21, 1957. A Flying Tiger DC-6 piloted by Capt. Gregory Thomas made a forced landing on a sand bar in Jamaica Bay after losing power in all four engines on takeoff from Idlewild. There were three minor injuries.

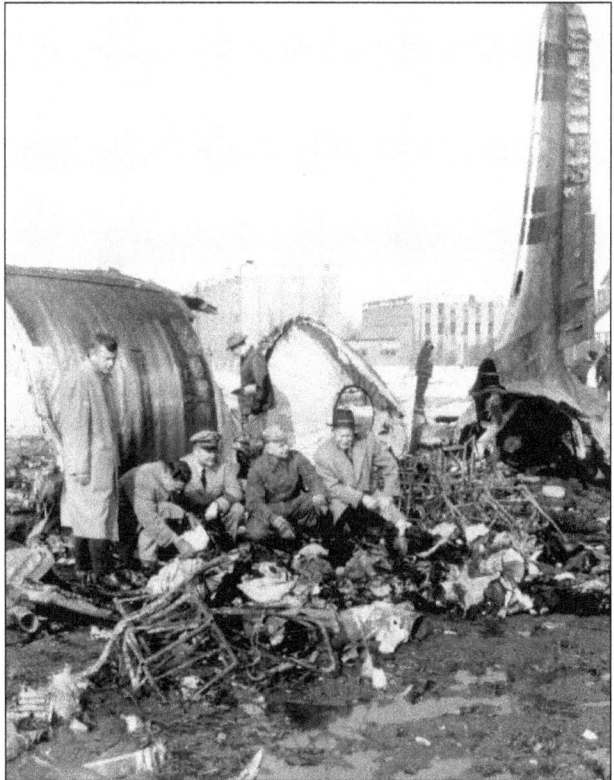

A FATAL DC-6 CRASH, RIKER'S ISLAND, JANUARY 2, 1957. A Northeast Airline DC-6 lost power one minute after takeoff from LaGuardia and crashed on nearby Riker's Island. In the crash, 22 people were killed and 35 people were injured.

An Eastern Airlines Lockheed Constellation. This aircraft attempted a forced landing at Islip MacArthur Airport on January 9, 1958.

The Fuselage of a Seaboard & Western Lockheed Constellation. On October 11, 1958, this aircraft collided with a Trans Canada Vickers Viscount on takeoff. There were no fatalities. Here, the fuselage is being trucked away at Idlewild Airport.

Six

TOWARD SAFER FLYING

From the late 1920s through the 1950s, Long Island was the center of the effort toward making flying safer. Due to the enormous number of air crashes, both locally and nationally, virtually any way of making aviation safer was carefully looked at. This effort was undertaken both by the government as well as by private industry. Thus, being a center of the aviation community, Long Island pioneered many aviation safety features whose legacies are still with us today. From developments in air-traffic control to new types of instruments to aircraft-safety equipment and state-of-the-art airports, Long Island was a pioneer. The many aircraft crashes that once littered Long Island ultimately led to aviation becoming one of the safest forms of transportation ever invented.

THE ADMINISTRATION BUILDING AND CONTROL TOWER. Floyd Bennett Field in Brooklyn, when opened in 1931, had the first control tower at a commercial airport in the United States. Control towers virtually eliminated the possibility of midair collisions in the vicinity of the airport.

The Headquarters Building and Control Tower, Mitchel Field. When this tower opened in 1936, this was the first control tower at a military airfield in the United States.

JIMMY DOOLITTLE IN THE CONSOLIDATED NY-2 HUSKY, MITCHEL FIELD, 1929. Using this modified aircraft with new instrumentation and a hood that covered the cockpit, Jimmy Doolittle made the world's first "blind flight" at Mitchel Field in September 1929. He took off, flew a course, and landed without ever seeing out of the covered cockpit. Using these new instruments developed on Long Island, aircraft could now fly safely at night and in bad weather for the first time.

A FORD TRIMOTOR COCKPIT, C. 1930. The Sperry Artificial Horizon, left center, was developed on Long Island in the late 1920s. For the first time, it allowed the pilot to accurately know the aircraft's attitude when he could not see the horizon.

122

INSTRUCTORS OF THE ROOSEVELT AVIATION SCHOOL, ROOSEVELT FIELD, 1932. Through the 1930s, the largest and most advanced flying school in the United States was the Roosevelt Aviation School at Roosevelt Field. Here, taught by professionals, students learned how to fly safely and carefully, much more so than in aviation's earlier days.

ROOSEVELT AVIATION SCHOOL NIGHT FLYING AIRCRAFT, c. 1930. When it began night flying instruction in 1929, the Roosevelt Aviation School was the first civilian school to do so. Here, instructors look at one of the new landing lights.

A LINK INSTRUMENT TRAINER, ROOSEVELT FIELD, C. 1932. When it began blind flying and instrument instruction in the early 1930s, the Aircraft Radio and Instrument Training Company at Roosevelt Field was among the first in the country to do so.

ROOSEVELT FIELD
CHOOSES SPERRY-A.G.A.
FLOODLIGHT

Quick to realize its many important advantages, officials of Roosevelt Field have selected the SPERRY—A.G.A. Floodlight.

Airport officials and pilots who have observed and used this light are one in praising its great power; sharp cut off, preventing blinding; its smooth operation; and reliability with its fully automatic lamp mechanism, eliminating the need of constant attention while in operation.

Will gladly supply detailed information concerning this remarkable Floodlight upon request.

SPERRY GYROSCOPE CO., INC.

BROOKLYN   NEW YORK

A SPERRY AIRFIELD LIGHTING SYSTEM ADVERTISEMENT, 1930. By 1930, Roosevelt Field was the first general aviation airfield in the United States to be fully illuminated for night flying.

An Aerial View of Roosevelt Field, Looking North, c. 1935. After its beginnings as a dirt and grass airfield, by the early 1930s, Roosevelt Field had the finest concrete runways of any general aviation airfield in America. Concrete runways eliminated the problem of aircraft nosing over upon hitting mud, water, a depression, or a soft spot in the dirt field.

PAVING THE RUNWAYS AT MITCHEL FIELD, 1932. Mitchel Field was also among the first military airfields in America to have long, wide concrete runways.

THE CURTISS TANAGER, 1930. Built in Garden City in 1929, the Tanager was the winner of the 1930 Guggenheim safe plane competition at Mitchel Field. This was the first plane to effectively incorporate wing flaps and slots, so aircraft now had good control at low speed while landing. These features were soon incorporated in new aircraft designs.

AN RCA FOKKER SUPER UNIVERSAL, ROOSEVELT FIELD, C. 1929. It was in this aircraft, based at Roosevelt Field, that RCA tested some of the first aircraft radios ever built.

127

THE LEARADIO LABORATORY, ROOSEVELT FIELD, C. 1937. Lear developed one of the first aircraft radio direction finding units (RDFs), which were first tested atop automobiles driving around Long Island. RDFs allowed aircraft to hone in on radio signals so they could accurately know their position at night and in bad weather.

AMELIA EARHART WITH THE NEW SPERRY GYROPILOT, BROOKLYN, C. 1935. Sperry's Gyropilot was the world's first automatic pilot produced. It allowed a specific aircraft course to be set and followed accurately and automatically.